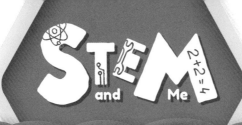

STEM
and Me
2+2=4

SENSES

By
Robin Twiddy

BookLife PUBLISHING

©2023
BookLife Publishing Ltd.
King's Lynn, Norfolk
PE30 4LS, UK

A catalogue record for this book is available from the British Library.

ISBN: 978-1-80155-825-9
ISBN: 978-1-80155-856-3

Written by:
Robin Twiddy

Edited by:
Hermione Redshaw

Designed by:
Amy Li

All facts, statistics, web addresses and URLs in this book were verified as valid and accurate at time of writing. No responsibility for any changes to external websites or references can be accepted by either the author or publisher.

IMAGE CREDITS

All images are courtesy of Shutterstock.com, unless otherwise specified. With thanks to Getty Images, Thinkstock Photo and iStockphoto.

Recurring images – dip, tonkhao wanpiya (patterns).
Cover – azure1, Coosh448, dotshock, G_O_S, Lizard, Yellow Cat, Kamira, N.Savranska. 1 – Kamira. 2–3 – azure1, G_O_S, Lizard, Yellow Cat. 4–5 – Deyan Georgiev, Russamee, VaLiza, svtdesign, Snezhana Togoi, Helen Nertis. 6–7 – svtdesign, janinajaak, Monkey Business Images, PixMarket. 8–9 – BNP Design Studio, svtdesign, graphic-line, Inna Vlasova, Littlekidmoment, Paysawat-Mizu. 10–11 – 3445128471, adecvatman, BNP Design Studio, Toey Toey, svtdesign. 12–13 – grey_and, Natalia Bostan, svtdesign, Oleksandr Rybitskiy, PinkPueblo, Pixel-Shot, Sergey Mastepanov. 14–15 – Studio Barcelona, svtdesign, Dusida, Ihor Bulyhin, Yulyazolotko, Sanit Fuangnakhon. 16–17 – Biscotto Design, Elena Yakusheva, Lio putra, Monkey Business Images. 18–19 – Helen Nertis, Colorfuel Studio, maradaisy, Monkey Business Images, Syda Productions. 20–21 – Colorfuel Studio, fizkes, Standret, svtdesign. 22–23 – fizkes, svtdesign. 24 – azure1, Lizard.

CONTENTS

Words that look like <u>this</u> can be found in the glossary on page 24.

WHAT ARE SENSES?

Our senses help us understand the world around us.

Humans have five senses. These are sight, hearing, taste, touch and smell.

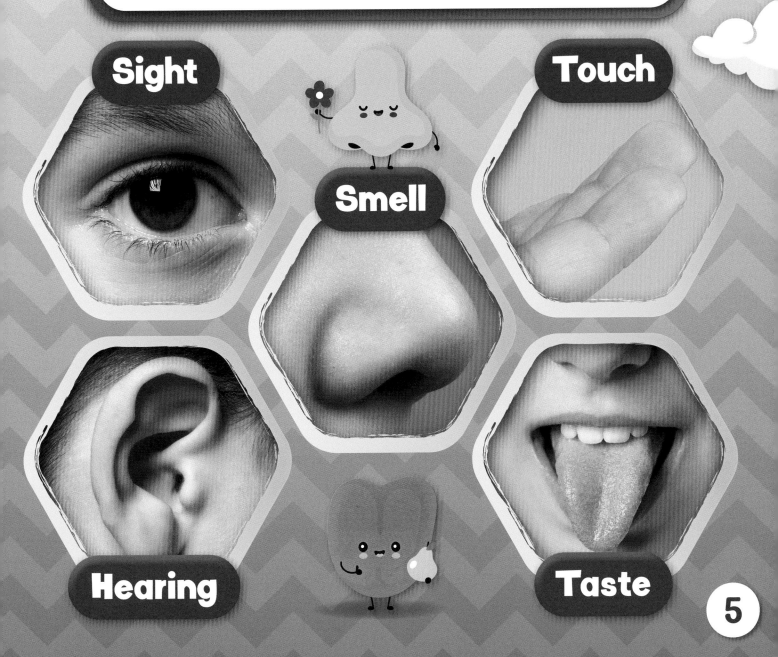

Sight

Smell

Touch

Hearing

Taste

5

SIGHT

Our eyes let us see the world around us.

We see colour, size, shape and more.

Our eyes need light to see. The darker it gets, the less we can see.

A torch helps us to see in the dark.

7

HEARING

We hear when sounds reach our ears.
We have an ear on each side of our heads.

We can hear things that are loud, quiet, close or far away.

Do you like listening to music?

TASTE

We use our sense of taste when we eat.

Is this watermelon sweet or <u>sour</u>?

Tongues have small bumps called taste buds. They tell us if food is sweet, sour or <u>bitter</u>.

TOUCH

Our sense of touch tells us about how things feel.

We can feel lots of different <u>textures</u>.

Soft

Smooth

What other textures can you feel?

Rough

13

SMELL

Our noses tell us how things smell.

Flowers can smell really nice.

14

Fresh bread has a strong smell, but so does bad cheese.

Yum

Yuck

What kinds of smells do you like?

SENSES WORKING TOGETHER

We don't just use one sense at a time. All of our senses work together.

There are lots of sights, sounds and smells at the farm.

When we eat, we see, smell, taste and hear our food.

An apple crunches when you bite it!

SENSES AT PLAY

Your senses are working when you play.

Which senses do you think you use when you play tag?

19

SENSE GAME

Take off your shoes and put on a blindfold. Now, have an adult turn you around three times slowly.

Have them walk you to a different room.

Now, try to listen, feel and smell.
Can you work out which room you are in?

Can you hear any sounds? What does the floor feel like?

Which sense helped you work out the room that you were in?

23

GLOSSARY

BITTER	a sharp, unpleasant taste that isn't salty or sour
SOUR	a tart or sharp taste
TEXTURES	the ways that surfaces look and feel

INDEX